Make Math

EASY

A Student's Guide to Succeed in Mathematics.

BOOKS

KARTI MITAL

First Printing: December 2008
08 09 10 11 12 13 14   10 9 8 7 6 5 4 3 2 1
Printed in the United States of America

Acknowledgments

Without certain people, this book would not be what it is. Though I would like to believe I know everything about this topic, there are still many words of advice that would not have been printed had it not been for some of the following individuals:

Mrs. Claire Kovar
Mrs. Heddi Sirovatka
Mr. Rick Cazzato
Dr. Arjun Gupta

They have helped me greatly in refining my concepts and giving me some practical suggestions. Throughout this process, they were eager guides.

In addition, I would like to thank my mom, dad, and brother for their many hours of motivation and careful proofreading. If not for you, I do not think this would have been possible.

Karti
Mital

Table of Contents

Karti
Mital

$$x^2 - \pi + x \div$$

INTRODUCTION

What's the Purpose of this Book?

If you quickly skim through this book, you'll see that there are hardly any numbers in it. Why is that? Well, this book is meant to teach you certain techniques. They range from easy time saving tricks to learning how to get an A on a test **before you take it**. Also, it will help you think like a smart person so that big scary problem suddenly becomes a lot easier. These techniques can help you do better in any math class, from Ms. Granger's Algebra I class to Mr. Studenbachen's Vector Calculus class. In this book, I will not teach you any specific math concept. Because of this, you will find that this book happens to be a lot shorter than most test prep books.

To put it into perspective, let's compare math to a tennis match. A lot of times, a player is as good as his opponent. He can hit ground strokes, volleys, and serves just as well as the other guy. The only reason he loses is because he hits to his opponent's forehand, which is strong, versus his backhand, which is weak. In this book, you will learn when you should hit to the opponent's backhand. In addition, it will teach you how to improve those shots in order to get ready for that match.

How does this book work?

This book will cover the classroom habits, your study time, and test taking strategies. I use very simple math concepts in my examples, so almost anyone who reads this will understand what is going on. Here is a quick overview of the different parts of the book.

The Classroom: You will soon find out that there is more to learning math than copying down what the teacher writes on the board. This section will reduce your study time and make sure you walk into the test better prepared.

Study Time: Yes, unfortunately, you will still have to study for math. After you read this

$$x^2 - \pi + x \div$$

book, though, you hopefully will get more out of it. In this part, you will learn what you should study in order to master the material. In addition, you will learn to think through those infuriatingly hard problems.

The Test: This is the part where I tell you all of those test taking tips that magically improve your grade. This section will allow you to get the most out of what you have learned.

After each section, there will be a couple of pages for notes for you to write down the ideas that resonate with you the most.

A Few Things About Math

Everyone (including you) can do math. Math takes many forms, from counting pennies, to finding the area of a circle, to finding the probability of a sunny day, and even to calculating the tangential velocity of a rocket. Your math class may seem hard to you, but when you think about it, there are millions of other people who have been in your seat. Even if you are in calculus you are not alone! Don't worry if you feel like you are overwhelmed by a certain math topic. Just remember that there are plenty of people who were once in your position. If those people got through it, so can you.

Your math grade does not tell how smart you are. Plenty of smart people just don't try in class, so they get a bad grade. On the other hand, some people who are not gifted in math work really hard and pull off an A. With the help of the tips in this book in addition to a little bit of work, you can get a better grade than you expected.

You really don't need to have the right answer. A lot of teachers will give you a generous amount of partial credit for having some of the work done correctly. Because of this, you can still get a decent grade even if you got only half of the problems right.

You are not bad at math. If you think you are terrible at math, then your grade won't be too pretty. If you put on a positive mindset, then you **will** do better no matter how hard the test is. Also, when you have trouble with something, don't get discouraged. You can do it. I cannot guarantee an A solely because of a positive mindset, but I am pretty sure you will get a bad grade if you think you will bomb the test.

$$x^2 - \pi + x \div$$

Chapter One

The Keys

If you are reading this book, you fit into one of two categories. You are either 1) really smart 2) average or 3) not really smart. If you are in the second or third categories, then get ready to learn how those smart people tick. If you're in the first category, then get ready for an ego massage.

Imagine you need to open a locked door. You have a key to open that door; the only problem is that you have about three million of them. How do you choose which one to use? Well, imagine if you could magically go through these keys really quickly. You might even be able to open the door before some other people get their keys ready! Also, imagine if you had this strange ability to know what key is needed for a certain lock just by looking at it. Other people may take forever just to find the right key, and some might not even have that key! Doesn't it feel great to be so good at opening

doors? Well, it's the same for being good at math.

The difference between a regular person and a smart person is that he or she just happens to be much faster. They can jump through mental steps and solve math problems (or open doors in the example) much more quickly than a regular kid. Because they are speedier, they do other things better. They grasp math concepts better (and thereby find the "key" to a door). They understand how to make huge leaps in logic to solve a really tough problem. These people also know when and where to use certain methods versus others. (For example, when the problem asks you to find the equation for a line, everyone knows the answer is not "Macbeth." A smart person, though, will be able to quickly decide which form he or she should use, such as point-slope versus slope intercept.)

But don't worry about that. The only real advantage that smart people have over others is their speed. All the other things they do better come from the fact that they can process information quickly. I'll share with you later some strategies to work with those keys more efficiently. So even if a smart person may be naturally faster than you, I can make the difference between your grade and his a lot smaller.

What does this book fix?

People can have a bunch of problems while taking a test. Some of these may look familiar to you. You will see pictures under the types of mistakes. When you see a picture later, you will know that it can help fix one of the problems mentioned below.

The duh…

Sometimes you walk into a test and get to the last problem. It asks you to do something, like find the square root of a number. You already learned how to do the rest of the stuff in the test, but you  didn't know that you had to learn that one thing. If you knew how to do it, the problem would be a cinch, but because you never learned the concept, you couldn't.

$x^2 - \pi + x \div$

The blank…

Once your teacher gives you the test, you forget everything you learned the past two weeks. Don't worry; there are ways to fix this.

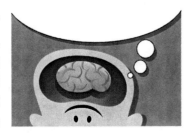

The curveball…

This is like a normal problem but with a little twist. Because of this one little change, you can't do the problem. For example, if you are learning slopes, and your teacher asks you to find the slope of a vertical line, then you might have a problem. Curveballs usually make sure

that you know the concepts well enough to deal with little nuances like the one mentioned above.

$$x^2 - \pi + x \div$$

The stupid mistake...

This seems to hit everyone at some time. Sometimes you do dumb stuff like write that 2+3=6. There are a few ways to help minimize this mistake.

The "Oh my God!"...

This kind of problem seems to be totally off in left field. It's kind of like the duh added to the curve ball. You have never even seen a problem like this before. In spite of this, the smart kids seem to

miraculously find the solution out of nowhere. Honors level teachers often put these problems in so that they can separate the A's from the rest of the class.

What Results Can I Expect?

I can't guarantee you an A from this book. But wait! Before you give up on this book, make sure you know a few things. This book can help you do better in class while also making you score more consistently on tests. Also, it will make sure you have a sound knowledge base that will help in more advanced levels of math. So, even if you don't do that much better this year, you at least will have a firm foundation for next year.

Notes

Karti
Mital

Chapter Two

Classroom Section

Ah the math classroom, the place where you listen to the boring drone of your math teacher telling you how to add fractions and work with percents. Some people like to sleep there, some people just take notes, some spend the whole hour asking questions, and a considerable amount just do their English homework.

If you use your time in class properly, you can spend less time later studying and doing homework. How? Well, as you soon will see, I will teach you how to (gasp!) actually learn something in class!

 ## Alertness

You probably want to learn in your math class. That means you don't want to be sleeping, daydreaming, talking to the person next to you, texting your best friend, or doing

your science homework. Trust me, if you actually pay attention in class, you will learn more and need less time to study. That will free up more time after school to do whatever you want. It might also prevent you from getting grounded, due to that F in math class.

- If you have trouble paying attention, there are a couple of possible cures. The first is to try to write down everything the teacher says. It may keep you from zoning out. It won't make your notes better, but you will at least stay on track.

- The second option is to take notes using different colors. This usually takes a lot of concentration, so it may keep you focused. This will not work if you like to doodle a lot.

- Finally, you can sit away from your friends or try not to talk to them. This may be hard to do and will not be fun, but if you really get distracted by them, then it will at least help your grade.

$$x^2 - \pi + x \div$$

Questions Please?

If you get lost, raise your hand and ask a question. There are plenty of kids who probably have no idea what's going on. To them, you will be their very outspoken hero. Also, by raising your hand, you may help slow down the class to a manageable pace. That way, your teacher won't try to stuff three chapters of vector calculus in your head all in one hour.

Don't worry about looking like an idiot. Your question may seem dumb to you, but it probably is not. If you ask that question in class, you will ask fewer questions later and be less confused with your homework. At your 10 year high school reunion, no one is going to remember you as the kid who always asked questions in math class. If they do remember your math class antics, they might just remember you as the kid who made the class go slowly enough to be bearable.

So let's go back to that concept that you don't understand. It will probably be needed later, so make sure that you understand it. It is important to know how and why your teacher does certain steps in practice problems. If the teacher happens to be explaining the first step in a very long process, and you do not understand this, then you will not follow what is going on afterwards.

For example, if your teacher is talking about slope, the first thing he or she might tell you is that slope is rise over run. If you do not understand what those words mean, then you probably will not understand what $(y_2-y_1)/(x_2-x_1)$ will mean.

If you still don't understand something, then you probably want to go in later and have him or her explain it to you one on one. If you keep on saying "I don't get it," then your teacher and the rest of the class probably will remember you as the idiot who made their class go way to slow, not as their savior.

For example, let's pretend that your teacher is explaining the difference between a line and a ray. Your teacher tells you that a ray has one endpoint and a line has no endpoints. You raise your hand and say "I don't get it." Your teacher might then say the same thing over, or draw them on the board to explain them. If you really do not get it after that, then it is probably best if you come in later to get help. Your teacher will be quite happy to explain it to you, it's just that he or she still has to teach the rest of the class.

Be Active

When your teacher asks for a volunteer to work through a problem in front of the class, do not be afraid to raise your hand. If you participate like this, then your teacher can see how well you understand the subject. If your teacher knows this, then he or she can decide if your class needs some extra time to go over a certain topic.

Let's say you are learning about how to add fractions. If your teacher sees that you do not know what the first step is at all, then she might be more inclined to push back the test a day so that she can explain it to you. I am pretty sure you want to see that happen more often.

In addition, if you participate, you are basically getting your own private tutor. Your teacher can guide you through the problem, make you think the right way, and make sure that you can actually solve a problem like this. If you participate in class, the help that your teacher gives you will almost certainly reduce your studying time.

So let's say that you are going over that fractions problem. You really have no idea where to start. If you raise your hand and try to answer it, your teacher can show how you need to first get a common denominator and then add the numerators. For raising your hand, your teacher has basically given you the one on one help that plenty of people shell out big bucks for.

Karti
Mital

Take Notes

OK, this does seem like a no brainer, but if you don't take the right kind of notes, then you are just wasting your time. So when your teacher explains a concept, make sure that you write down what is said **in your own words**. Plenty of teachers use power points or  overheads to write out what they want to explain to you. If you just copy that down, and you try to study the material later, you will probably end up wasting ten to fifteen minutes deciphering what the teacher meant.

Let's pretend that your teacher tells you that a hypotenuse is the "square root of the sum of the squares of the two legs." Since you don't regularly speak in complex math jargon, you might want to translate that into English. To make that simpler, you could say "to find the hypotenuse, square each of the smaller sides, add them together, and then find the square root of that answer." My paraphrase may not make sense to you either, which is why when you take notes, make sure that **you** can understand them. Remember, they are **your** notes. This means that you have to raise your hand and ask your teacher what the concept means if you are confused.

In addition to writing concepts in your own words, you should also explain formulas. If your teacher gives you the formula $a^2 + b^2 = c^2$, then you might want to do the following:

- Write the name of the formula (so that you know what your teacher is talking about).

- Write down what the formula is used for (in this case finding the hypotenuse of a triangle).

- Write down if the formula has any limitations (In this case, the triangle must be a right triangle).

- Write down what the variables mean (a and b are the legs and c is the hypotenuse).

If you feel like doing this slightly differently (like writing these things in a different order) feel free to do so. Remember, the notes that you write are for **you**, so they should make sense to you and follow the format that you like the most. But make sure that you write the

$$x^2 - \pi + x \div$$

important stuff down so that you don't run into a problem on the test.

Homework

If you have trouble with your homework, make sure that you go in to see your teacher. You should first probably check with a parent or a friend. If it seems to make sense to no one, then it will help your understanding of the material and your grade if you go in for help.

Often, a teacher will assign problems that have answers in the back of the book, so make sure you check them. I have some friends who think they are ready because they did their homework, but are not because they did not check it with the answer key. So when they go to take a test, they do a couple of steps incorrectly, but do not know about it. Because of that, they can get a bad grade just because they were too lazy to use the back of the book to check their answers.

$$x^2 - \pi + x \div$$

What's on the Test?

It always helps to ask your teacher what will be on the upcoming test. Some will be nice enough to tell you beforehand that the test will cover all of the material in the chapter, but some will not. If your teacher skips around a lot in the book, make sure you ask because there is a chance that something completely random will pop up. When asking, make sure you write down what your teacher  says, or chances are that you will forget what he or she said. This will obviously reduce the time that you have to waste on calling friends to see what is on the test. You will also avoid the duh! questions on the test. In addition, ask the teacher how much time you should spend on each type of problem. This will help you with the easy questions on the test. In short…

- Ask for what is on the test
- Write it down

- Ask how much time each problem should take

So in case you were completely comatose (unconscious) while reading the above section, in a classroom make sure to:

- Pay attention
- Ask Questions
- Participate, it's a free tutor session
- Take notes in your own words
- Come in for homework help
- Ask about what is on the test

Notes

Karti
Mital

$x^2 - \pi + x \div$

Chapter Three

Study Section

Studying. Most people really do not like to do it, but everyone knows that you have to study at least a little to do well. Some people study every night for an hour, some cram it all in the night before, and some people try to learn the material in the passing period before the test.

In this section I am going to help you study more efficiently. For some of you, that means more studying. For most, that means a lot less studying. Right now, I am going to make sure that when you study, you actually prepare correctly for the test.

Just a Little Background

When you study, you most likely are preparing for a test. Though that seems pretty obvious, plenty of people waste their time either studying the wrong stuff, or studying the wrong

way. Here are a few things to know before preparing for a test.

- Your math teacher is not just checking to make sure that you know the concepts. Otherwise your math test would consist of just writing down terms, formulas, and definitions. If you do not do practice problems, then you are like a scared sixteen-year-old who has read about driving, but has never actually sat in the driver's seat. That can get a little scary.

- There are some problems that your teacher has never explicitly shown you. Most of these are curveballs and will require you to figure out a way to solve a problem with what you know. You will probably never go over this type of problem in class because your teacher wants to differentiate the A's from the B's. Even if you can't get all of them right, there are some ways to ensure that you can control how you do on them. I will show you some ways in Step 3: Thinking and in the test taking section.

- For higher and honors level courses, there are some problems that do not have a certain method that can be used to solve them. These include geometric proofs, trig identities, and integrals to name a few. In order to do well at these, you need to practice. Even if you are extremely intelligent, if you have not done a few of these before the test, then you are going to bomb the test.

Step 1: Starting Out

At this point, you should look over the terms, concepts, and formulas. Make sure you know what all of them mean. You could be doing this step right after the teacher has explained the material to you. Most likely, though, you should do this in the evening while you are doing your homework. If you do this little bit of preliminary studying (this takes all of thirty seconds), you can save time later and also have a better grasp of the subject. Also, try to memorize the material a little bit to improve your grasp of the topic.

- When doing your preliminary studying, you should be able to come to a certain level of understanding with the topic. You should know **how** and **why** the concept works a certain way. With formulas, you should know how they are derived and why they work, unless that is beyond the scope of your math class.

- For example: If your teacher is teaching you direct variation, he or she may give you the simple formula $y = kx$. Instead of just memorizing the formula, you should be able to see that when x increases, y has to increase (since k is a positive constant). That way, if you forget the formula on a test, you could have a way of figuring it out by working with the definition. Also, if you somehow used the wrong formula, you could see that your result does not make sense.

- When doing this preliminary step, it may also help to understand the concept visually. Some people are not visual learners, so this step may not be necessary, but it will still help create a deeper level of understanding. For example, with inverse variation, it might help to use the see-saw analogy. As one side, or x, increases, the other side, y, will go down. Remember, some things may be harder to understand visually, so a visual will not always help.

Step 2: Practice

When studying for a test, you **will** have to do practice problems. What most people do not know is that there is a way to save time doing them before a test. It's called homework. When you do your homework, you are really just studying for the test. While doing this, make sure that you understand why things work the way they do. Homework is a great way to see if there are any holes in your understanding of a concept.

For example, if you have a problem that says "What is the slope of the line with the points (0,0) and (0,5)." When you work through it, you come to the answer of 5/0, so you write down 0 as your answer. When you look at the answer in the back of the book, it says "und." You have no idea what "und." means, so you check the front of the book where it explains the concepts. There they tell you that a slope with a denominator of 0 is undefined, answering your confusion. This way, you just possibly avoided a curveball on the test. This definitely takes more time, but it pays off on the test.

- If you do not understand a problem, do not give up and just write a random number. Ask your parents if they can help you with it, or see if you teacher can

help the next day. There is a good chance that the problem may be on the test, so if you get some help, it will help you later on.

 • Some teachers rarely check the homework or do so at the end of the unit. Do not try and do it all the night before if you are in this situation. If you do it following the schedule that he or she gives, you can understand the material better and not fall behind in class. The teacher has assigned the homework in this order so that you follow the class schedule and actually learn the material.

 • Think of things differently. If you can think of a certain concept in another way (such as with geometry) it may help you get that full understanding to help you fend off curveballs or the Oh my god's.

 • When you practice, improve your speed. If you can do easy problems very quickly, it will save you time on the test for the hard ones.

Step 3: Thinking

The difficult ones: When you encounter a difficult problem, imagine that you are at that locked door. There can only be two reasons why you cannot solve the problem. You either can't find that key, or you just do not have the key. If the key is lying around, you might have to turn it a bit more than you thought get the correct

answer. In order to find a way to do the problem, there are a few questions that may help you out. They follow the acronym QASSD.

 i. **What is this <u>Q</u>uestion asking?** See what your teacher is asking for. Oftentimes, a teacher will use a lot of difficult words to ask for a very simple thing. Once you know what the question is asking, then move on to the next step. For example, if your teacher asks you to "ascertain the standard form equation of the line that is normal to the line defined by the following points: (0,9) and (1,17)," then try to make it simpler. They basically are asking for you to find the slope of a line. This line is normal, which means it is perpendicular to (or makes a right angle with) the other line that has those two points.

ii. **What do I know that has <u>A</u>nything to do with this topic?** If you don't know where to start on a certain problem, start thinking about concepts that are kind of related. So, if they ask for the equation of the line in point slope form, then see if you know what makes up point slope form. Then see if you can find those things.

iii. **How far can I go in this problem before I get <u>S</u>tuck?** Pretend that you have the two points, but you don't know how to get the slope. You know the setup of point slope $(y-y_1 = slope*(x-x_1))$. At this point, you have two options. 1) Write down the equations using one point and a made up slope 2) Try to find the slope. In order to do this step, you need to dig deep into your brain.

iv. **Have I done anything <u>S</u>imilar to this?**
It may help to remember how you make other forms, such as slope intercept form. This can sometimes jog your memory so that you remember what to do. It does not work all of the time, though, so do not waste too much time on this step.

v. **Can I think about this in a <u>D</u>ifferent way?**
You might want to try thinking about the question in a different light. For example, you could think about what a line is. It is a set of points. The equation finds these points by using a slope. You now know that the slope tells you the difference between the two points.

vi. **If these questions do not work**…
Then you probably want to work on your knowledge base. There is probably something (this something is probably very small) that you did not learn. It might be a duh! kind of problem, or you might be missing out on an intermediate step. Either way, make sure that when you try to open that door, you have all the keys you need.

Shortcuts: When you are studying, if there is a quick and easy way to do something, then try to learn it. The shortcuts that you come up with work best. Sometimes you find a way to save some steps, sometimes you think of a certain process in a way that makes more sense to you. Though this makes more sense to you, it may not to others. That is the beauty of a shortcut. Your teacher will teach you something in the way that is easiest for the majority. If something works out faster for you, then you are in luck.

Note: Though shortcuts help, make sure that you know how and why it works. A lot of times, people use shortcuts that do not work all of the time. Then the teacher gives out a slightly different question, and half the class gets it wrong. When using a shortcut, treat it like a formula; know where it's used, what it does, and what its limitations are. If you are unsure about these, ask your teacher.

Different Approaches: If you have trouble grasping a concept, and no matter how many times your teacher explains it to you it still does not make sense, then you might want to try a different method. Here are a few...

$$x^2 - \pi + x \div$$

- **Try to learn the concept visually**. I explained this earlier in Step 1, but may help out if you still have trouble with a concept. Almost anything in math can be represented visually, so this is a method that can work if you are stuck.

- **Think of it as a mystery**. A lot of times in algebra, you are solving for a certain variable. Pretend that what you are looking for is a puzzle and try to solve it. Your mindset is a big part of math, so it often helps to think of math as something else. This may help you work through seemingly impossible concepts.

- **Have another person explain it**. Older siblings and friends who have already gone through the material are often good people to go to if you have trouble. If they are good explainers, chances are that they can explain it to you in language that is familiar. Also, they may have a different approach to the material that the teacher does not. These people are often willing and very able to help you out.

$$x^2 - \pi + x \div$$

- **Go online**. There are plenty of websites that will explain math concepts in plain language. If you are ever stuck on some problems, I have some references in the back of this book just for that purpose. Try them out and see if they work for you.

Step 4: The Night Before

By this time, you should have a good grasp of what is going on. Now just make sure that you memorize and fully understand the concepts, terms, and formulas for the unit. Once you have everything fully memorized, go through each type of problem at least once. Make sure that you can do them all by yourself. Also, make sure that you can do them quickly (ask your teacher how fast you should be able to do it). Be certain that you can do the problems that you had trouble with before because they probably will find their way onto the test. In short, you should

- Have everything memorized
- Be able to do the problems alone
- Be able to do them quickly
- Concentrate on the difficult topics
- Not spend too much time on practice problems if it cuts into your sleep

"Reading" vs. "Writing": Math is like driving. You can't really do it just by watching others. When you are doing problems, it is very important that you can do them without the help of your book, notes, parents, or friends. You **need** to do the problem all **by yourself**. If you have someone tell you how to do even a little step of a problem, chances are that you really do not know what is going on.

For example, imagine that you are taking a spelling test. Most people are able to read the words on the list with no problem. In math, if a teacher goes through a problem with you, chances are that you sort of understand what he or she is doing. But in order to get an A on a spelling test, you have to be able to write out the words yourself. Though you may have been able to read it earlier and spell most of the word correctly, it does not matter. You can still get that word wrong. The same is for a math test. Even if you understand Mr. Finklestein's explanation of how to solve the problem, you are not guaranteed an A on the test. You still need to be able to do the problem by yourself. So, just to make sure you can "spell", go over all of those problems with no one (that includes your notes) around you to see if you can truly "write" out the answer.

Karti
Mital

Remember: While studying, make sure that you:

- Know how or why the concepts work
- Don't give up on problems
- Try to get fast at doing the problems
- Make sure you are reading and writing
- Remember QASSD when you are stuck
- Try out some shortcuts

Notes

Karti
Mital

Chapter Four

Test-Taking

Plenty of people freak out about tests such as the SAT or the ACT. Plenty of people buy books from companies with big fancy names to prepare for those tests. They seem to have a few tricks to make you test taking experience a lot easier. Well what if there was a book that could do the same for your ordinary math test. What if you were reading it right now...

If you haven't figured it out from the title of this section yet, this is the part where I share some tips on how to be a good test taker. I am going to show you a few tips and teach you a few strategies that help. Since you are not me, some of these strategies may not work for you. When you come across these, I will put a little disclaimer. Just look for the exclamation mark; they tell a process that may not work for all.

After you read this section, make sure you practice these strategies. If you try to take a test without ever having used these methods,

then you probably won't like the result. Make sure to go through a test once with these strategies so that you are comfortable with them.

Sleep and Eat

It is very important that you get a full night's sleep and that you eat a good breakfast. If you have to study late into the night, make sure to sleep after a certain point. Math is one of those subjects where you really do not benefit that much from studying for hours on end. When studying late, make sure that you at least know the concepts, terms, and formulas. Do not spend a lot of time doing practice problems if it will limit the amount of time that you sleep.

The Blank

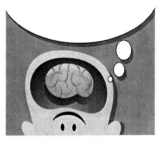

For those who forget everything the minute they get the test, make sure you write down everything you need to know. You can do it on the top of the test, or you can ask your teacher for a separate piece of paper after you get the test. For this sheet:

$$x^2 - \pi + x \div$$

- Do not take too much time because then you will put yourself at a disadvantage.

- You should have already planned out what you are going to write on it.

- Only write down things that you think you will forget. You are not going to forget the underlying concept that everything is based on (hopefully).

- This sheet is good for formulas, so write them down if you have trouble remembering them.

The First Run

In order to do well on the test, you will need to check your work. Before doing this, make sure you go through all of the problems in a first run. Here is what you should take note of

- **Speed**: Your teacher expects you to go slowly when taking a math test. This doesn't mean that you can't go faster than expected. Try to go as fast as possible, making sure that you are thinking properly. You should have practiced doing the problems in your homework quickly already. When you work quickly on the easy problems it gives you more time for the hard ones. If you somehow mess up on a problem, your speed will give you time to come back to it later in order to fix it To put it into perspective, when you are running out of time on a test, you often move at an incredible rate. Try to do 70-80% of that for the whole test. By the end of the test, you should be a little tired. It may be hard to keep up that kind of pace at first, but if you practice doing it all of the time, you will be fine.

- **Skipping Questions**: There are certain situations where you should skip a problem. Here they are:

  o If you get to a hard question that you can't do, skip it. Do not waste time trying to figure out how to start it. This wastes time that could be spent doing two or three easy questions.

  o When you are running out of time on a multiple choice test, skip the big problems. Guess an answer (if there is no guessing penalty) and go to the shorter ones. This way, you can maximize the number of questions that you get correct and don't waste time.

o In a free response test (which is what most math tests are), do not skip big questions. These usually account for more points than the small ones, so skipping them will not really give you more points. Only skip it if it you don't know how to do it or if your teacher gives short problems a lot of credit.

- **The Calculator**:
- (If you do not use a graphing calculator, then this section will not help you.) When taking a test, make sure that you save time by using your calculator intelligently. When you are doing intermediate steps or get an answer with a very long decimal, store it as a variable. This way, when you are checking your work, you

can just plug in that number. Otherwise, you have to do all of your steps over again. This is the same for functions. If you need to graph something like $y = x^3 + 5x$, do not erase that function. Instead, turn that function off so you can use it later. If you don't know how to do this, ask you teacher, or learn from someone who is good with calculators.

- **Your Teacher**: Your teacher can help you greatly while you take a test. If you are unsure about what a question is asking for, go up to your teacher and ask what it means. Also, if you encounter a hard question that you have no idea how to start, asking your teacher about a part of it may help you understand a way to tackle it. Some teachers are also nice enough to give hints. Watch out, though. If you are just fishing for answers, your teacher may not let you ask any more questions, which could hurt you if the teacher made a mistake on the test. In all, ask questions when you are confused, but don't try to wring an answer out of the teacher.

The Second Run

Once you are done with your first run, make sure to go back and look at the questions that you skipped. You may have to do this step a couple of times if you have trouble with the hard ones.

- If you skipped the problem because you thought you would not have enough time, make sure you do it.

- If the problem was too hard for you, then see if you found a way to do it. If you still have no idea, then try to write down the setup, or go as far as you can. This way, you can at least get partial credit. Unless you **really** are in a time crunch, **never leave a question blank**. Even if you put only half of the work, the first steps that you wrote out show your teacher that you at least know something about the problem.

The Third Run (AKA Checking)

No matter how well you think you have done on test, if you do not check your work, you will see a fair amount of stupid mistakes. By doing some of the following, you can

significantly lower the number of dumb things that you do. Just remember the acronym, RAPS.

- **<u>Re</u>-read the question**: Make sure that you did not do the problem incorrectly. For example, if the question asks for a line in slope intercept form, check that you do not have it in point slope form. This is also the time to make sure that you have the right setup. Mistakes with this part usually come from using the wrong formula.

- **Check your <u>a</u>rithmetic**: Go through the steps of the problem to see if you added something wrong or if you made an algebra mistake. If you are allowed to, do the steps in a calculator to make sure that you did not do something wrong in your head.

- **<u>Plug</u> it in**: If your answer solves an equation, then plug it in to see if the value you got is correct. This should be the first step in problems where you are solving for a certain value. If you have stored the answer in you graphing calculator, then you will be able to save time plugging it into the equation.

- o For example, if you are solving the equation $x^2 + 2x + 1 = 0$, put the value that you get as an answer into your calculator for x to see if it equals 0.

- **Does it make <u>s</u>ense?**: See if your answer makes sense. If you have an equation for a line, and your slope is 5 billion, then either your teacher is messing around with you, or you really messed up.

- **I know it is right**: Make sure to check all of your work. Even if you think that your answer is right, plug it in or go through your steps quickly. You might have made a small mistake that throws your entire answer off.

- **When you're done**: Before you hand in your test, make sure to have one last look at anything you have skipped. You may think of something new to help answer these questions.

Karti
Mital

Notes

Chapter Five

Quick Facts
to
Remember

If you want to look at this book later for quick reference, do not forget the following.

Remember, you can do math. So while…

In class

- Pay attention
- Ask questions
- Participate, it's a free tutor session
- Take notes in your own words
- Come in for homework help
- Ask about what is on the test

Studying

- Know how or why the concepts work
- Don't give up on problems
- Try to get fast at doing the problems
- Make sure you are reading and writing
- Remember QASSD when you are stuck
- Try out some shortcuts

Taking a Test

- Get a good night's sleep
- Run through your test at least three times
- Work quickly
- Skip questions strategically, but never leave a question blank
- Ask your teacher for help
- Check your work (remember RAPS)

$$x^2 - \pi + x \div$$

Notes

Karti
Mital

$x^2 - \pi + x \div$

References

All Subjects:

Math.com:
(This teaches concepts for all sorts of topics)
http://www.math.com/

Hotmath:
(This has step-by-step answers for almost every math book) http://hotmath.com/

Algebra 1:
Coolmath Algebra (This teaches the concepts)
http://www.coolmath.com/algebra/Algebra1/index.html

Yahoo Education:
(This has numerous practice problems)
http://education.yahoo.com/homework_help/math_help/algebra1

Geometry:
Free Math Help.com (Teaches the concepts and has some videos)
http://www.freemathhelp.com/geometry.html

Algebra 2:
Coolmath Algebra (This teaches the concepts)
http://www.coolmath.com/algebra/Algebra2/index.html

Yahoo Education (This has numerous practice problems)
http://education.yahoo.com/homework_help/math_help/algebra2

Trigonometry:
S.O.S. Math (This explains the concepts)
http://www.sosmath.com/trig/trig.html

Precalculus:
Coolmath Algebra (This teaches the concepts)
http://www.coolmath.com/algebra/PreCalc/index.html

Calculus:
Karl's Calculus Tutor (This teaches Calculus by using examples from everyday life)
http://www.karlscalculus.org/calculus.html#s4

Notes

$$x^2 - \pi + x \div$$

Reviews

Karti Mital has written a book that will benefit any math student regardless of one's ability level. The text is engaging, easy to read, and contains suggestions and strategies that are immediately applicable to a student's studies. Hard work is always necessary for one to achieve success, but Mr. Mital has provided a plan and eliminated the mystery that tends to shroud that road to achievement.

-Heddi Sirovatka
Algebra I, Honors Algebra II/Trigonometry, and AP BC Calculus Teacher

Math anxiety and Math phobia are not so uncommon to many of us. It is a feeling of frustration or helplessness about one's ability to do math. Karti Mital understands the problems and difficulties of learning mathematics since he has the first hand knowledge. He identified the need and has fulfilled it by providing this book that reflects his own experience during his

school years. Therefore in my opinion, this book should be of help to many students who have to take mathematics courses to meet the school/ state graduation requirements.

Arjun K. Gupta, Ph.D.
Distinguished University Professor &
Professor of Mathematics and Statistics
Bowling Green State University,

    Great Advice!

Claire Kovar
4th Grade Teacher

Author Bio

Karti Mital is currently a student at Hinsdale South High School, where he is ranked first in his class. Among numerous awards, he was part of the 1st place Freshman team for the ICTM Math Competition. In addition, he is also a member of the Varsity Scholastic Bowl Team, Science Olympiad, Model UN Club, Varsity Tennis, Investment Club, National Honor Society, Investment Club, and the Jazz Band at his school.

$$x^2 - \pi + x \div$$

Karti Mital is available for speaking engagements and personal appearances. For more information contact:

Karti Mital
C/O Advantage Books
P.O. Box 160847
Altamonte Springs, Florida 32716

To purchase additional copies of this book or other books published by Advantage Books call our toll free order number at:
1-888-383-3110 (Book Orders Only)

or visit our bookstore website at:
www.advbookstore.com

Longwood, Florida, USA
"we bring dreams to life"™
www.advbooks.com

Printed in the United States
211959BV00004B/1/P

9 781597 551731